Oliver J. Lodge

Some Recent Thought-Transference Experiments

Elibron Classics
www.elibron.com

Elibron Classics series.

ISBN 0-543-70413-0 (paperback)
ISBN 0-543-70412-2 (hardcover)

This Elibron Classics Replica Edition is an unabridged facsimile of the edition published in 1892.

From the PROCEEDINGS OF THE SOCIETY FOR PSYCHICAL RESEARCH,
Part XX., 1892.

SOME RECENT THOUGHT-TRANSFERENCE EXPERIMENTS.

BY PROFESSOR OLIVER J. LODGE, F.R.S.

IV.

SOME RECENT THOUGHT-TRANSFERENCE EXPERIMENTS.

By Oliver J. Lodge.

While staying for a fortnight this autumn in the house of Herr von Lyro, at Portschach am See, Carinthia, I found that his two adult daughters were adepts in the so-called "willing-game," and were accustomed to entertain their friends by the speed and certainty with which they could perform actions decided on by the company; the operator being led either by one or by two others, and preferring to be led by someone to whom she was accustomed. Another lady staying in the house was said to be able to do things equally well, but not without nervous prostration.

On the evening when I witnessed the occurrences nothing done could be regarded as conclusive against muscle-reading, though the speed and accuracy with which the willed action was performed exceeded any muscle-reading that I had previously seen, and left me little doubt but that there was some genuine thought-transference power.

Accordingly I obtained permission to experiment in a more satisfactory manner, and on several occasions tested the power of the two sisters, using one as agent and the other as percipient alternately. Once or twice a stranger was asked to act as agent, but without success.

The operations were conducted in an ordinary simple manner. One of the sisters was placed behind a drawing board, erected by me on a temporary sort of easel, while the other sat in front of the same board; and the objects or drawings to be guessed were placed on a ledge in front of it, in full view of the one and completely hidden from the other.

Naturally I attended to the absence of mirrors and all such obvious physical complications. The percipient preferred to be blindfolded, but no precaution was taken with reference to this blindfolding. Agent and percipient were within reach of one another, and usually held each other's hands across a small table. The kind and amount of contact was under control, and was sometimes broken altogether, as is subsequently related.

The ladies were interested in the subject, and were perfectly willing to try any change of conditions that I suggested, and my hope was gradually to secure the phenomenon without contact of any kind, as I had done in a previous case (Vol. II., p. 189); but unfortunately in the present instance contact seemed essential to the transfer. Very slight contact was sufficient, for instance through the backs of the knuckles; but directly the hands were separated, even though but a quarter of an inch, the phenomena ceased, to reappear again directly contact was established. I tried whether I could bridge over the gap effectively with my own, or another lady's, hand; but that did not do. I also once tried both sisters blindfolded, and holding each other by one hand, while two other persons completed the chain and tried to act as agents. After a time the sisters were asked to draw, simultaneously and independently, what they had "seen"; but though the two drawings were close imitations of each other, they bore no likeness to the object on which the agents had been gazing. (See Plates at the end of this paper, Plate I., R. 1 (*a*) and R. 1 (*b*).) My impression, therefore, is that there is some kind of close sympathetic connexion between the sisters, so that an idea may, as it were, reverberate between their minds when their hands touch, but that they are only faintly, if at all, susceptible to the influence of outside persons.

Whether the importance of contact in this case depends upon the fact that it is the condition to which they have always been accustomed, or whether it is a really effective aid, I am not sure. I have explained to them the desirability of securing the results without contact, if possible, and it may happen that with some practice they will find themselves able to dispense with it. The time at our joint disposal was limited, and I was not able to do more than set things in train for more striking success later, provided they try experiments occasionally and do not get tired of them.

So far as my own observation went, it was interesting and new to me to see how clearly the effect seemed to depend on contact, and how abruptly it ceased when contact was broken. While guessing through a pack of cards, for instance, rapidly and continuously, I sometimes allowed contact and sometimes stopped it; and the guesses changed, from frequently correct to quite wild, directly the knuckles or finger tips, or any part of the skin of the two hands, ceased to touch. It was almost like breaking an electric circuit. At the same time, partial contact seemed less effective than a thorough hand grasp.

It is perfectly obvious how strongly this dependence on contact suggests the idea of a code; and I have to admit at once that this flaw prevents this series of observations from having any value as a test case, or as establishing *de novo* the existence of the genuine power. My paper only appeals to those who, on other grounds, have accepted the general possibility of thought-transference, and who, therefore, need not feel unduly strained when asked to credit my assertion that unfair practices were extremely unlikely; and that, apart from this moral conviction, there was a sufficient amount of internal evidence derived from the facts themselves to satisfy me that no code was used. The internal evidence of which I am thinking was : (1) the occasionally successful reproduction of nameless drawings ; (2) the occasional failure to get any clue to an object or drawing with a perfectly simple and easily telegraphed name ; (3) the speed with which the guesses were often made.

I wish, however, distinctly to say that none of the evidence which I can offer against a prearranged code is scientifically and impersonally conclusive, nor could it be accepted as of sufficient weight by a sceptic on the whole subject. It is only because, with full opportunity of forming a judgment, and in the light of my former experience, I am myself satisfied that what I observed was an instance of genuine sympathetic or syntonic communication, and because such cases seem at the present time to be rather rare, that I make this brief report on the circumstances.

I detected no well-marked difference between the powers of the two sisters, and it will be understood that one of them was acting as agent and the other as percipient in each case. Sometimes the parents of the girls were present, but often only one or two friends of my own, who were good enough to invite the young ladies to their sitting-room for the purpose of experiment; though such experiments are, when carefully performed, confessedly rather tedious and dull.

In the early willing-game experiments, such things were done as taking a particular ring from one person's hand and putting it on another's ; selecting a definite piece of music from a pile, taking it to the piano, and beginning to play it. The last item (the beginning to play) I did not happen to witness, but I was told of it by several persons as more than could be accounted for by muscle-reading. A sceptic, however, could of course object that imperfect bandaging would easily enable a title to be read.

One of the things I suggested was aimed at excluding the operation of unconscious muscular guidance as far as possible, and it consisted in desiring that the lady while standing in the middle of the room should kick off her shoes without touching them and begin to sing a specified song. Success, however, was only partial. After one or two attempts to wander about the room as usual, she did shuffle a shoe off, but though she did not actually touch her feet she stooped so that the held hand came very near them. She then stood some little time uncertain what to do next, and at last broke silence by saying "Shall I sing ?"

The first attempt at the more careful experiments was not at all successful, but novelty of conditions may fairly be held responsible for that. On the second and subsequent evenings success was much more frequent: on the whole, I think, more frequent than failure. I proceed to give a fairly complete account of the whole series.

The first object was a teapot; but there was no result.

The first drawing was the outline of a box with a flag at one corner; but that produced no impression.

Next, for simplicity, I explained that the object this time was a letter (Buchstabe), on which it was correctly guessed E. Another letter, M, was given quite wrong. A childish back-view outline of a cat was given oval like an egg; some other things were unperceived.

On the second evening I began by saying that the object was a colour; on which red was instantly and correctly stated.

A blue object which followed was guessed wrong.

An outline figure of a horse was correctly named. So was the letter B. I then drew a square with a diagonal cross in it and a round ring or spot just above the cross, the whole looking something like the back of an envelope. After a certain interval of silence (perhaps two minutes) the lady said she was ready to draw what she had "seen," and drew the thing almost exactly, except that the dot was put right on the centre of the cross instead of above it, and a superfluous faint vertical stroke was added. (See Plate I., O. 2 and R. 2.) Its possible resemblance to an envelope was not detected, nor did the reproduction suggest the idea: it was drawn as, and looked like, a nameless geometrical figure.

The reproductions were nearly always much smaller in size than the originals. The agent did not look on while the reproduction was being made. It is best for no one to look on while the percipient draws, to avoid the possibility of unconscious indications. The original drawings were always made by me, sometimes before, sometimes during the sitting. These conditions were all satisfactory.

On the third evening I began with a pack of cards, running through them quickly; with 2 reporters, one recording the card held up, the other recording the guess made, without knowing whether it was right or wrong. I held up the cards one after the other and gave no indication whether the guesses were right or wrong. The suit was not attempted, so that the chances of error were, I suppose, 12 to 1.

On comparing the two lists afterwa. ds, out of 16 guesses only 6 were wrong. Full contact was allowed during this series. The lists are reproduced below.

The card guessing is obviously not of the slightest use unless *bona fides* are certain, but, given that, it affords the readiest method of studying the effect of varied conditions, interposed obstacles, and such like. The whole pack was always used and I simply cut it at random and held up the bottom card. About 10 or 12 cards could be got through in a minute.

The following is the list of the first card series. Full contact allowed:—

CARD LOOKED AT.	CARD GUESSED.
Seven of Spades	Seven
Six of Hearts	Six
Queen of Spades	King
Nine of Spades	Nine
Three of Spades	Six
Eight of Diamonds	Eight
Ace of Clubs	Ace
Knave of Diamonds	Queen
Five of Diamonds	Five
Two of Spades	Ace
Ten of Hearts	Six
King of Diamonds	King
Ace of Spades	Ace
Nine of Diamonds	Six
Eight of Hearts	Eight
Four of Spades	Four

Thus, out of the sixteen trials, 10 were correct and 6 were wrong.

Some guesses were made, both with cards and objects, on another evening, without contact, but none were successful. With contact there was success again.

I then went back to simple drawings; with the result that a cross was reproduced as a cross (Plate II., O. 3 and R. 3); a figure like 4 petals was reproduced in two ways, one of them being a vague 5-petalled figure. (Plate II., O. 4 and R. 4.)

An object consisting of an ivory measure standing on end like an inverted V was drawn as shown in the diagram, Plate II., R. 5 (*a*) and R. 5 (*b*).

A sinuous line was reproduced as a number of sinuous lines (Plate III., O. 6 and R. 6); a triangle or wedge, point downward, was reproduced imperfectly. (Plate III., O. 7 and R. 7.)

On other evenings more of the same sort of thing were tried, such as the diagram marked O. 8, reproduced as R. 8 (*a*) and R. 8 (*b*) (Plate IV.); a face, reproduced as 3 rounds with dots and cross (Plate IV., O. 9 and R. 9); and a figure like an A with an extra long cross stroke, which could be easily signalled as an A, but which was reproduced correctly as a geometrical diagram with the long stroke prominent. (Plate V., O. 10 and R. 10 (*a*), (*b*) and (*c*).)

A circle with 3 radii reproduced as a circle with roughly inscribed triangle. (Plate V., O. 11 and R. 11.)

The number 3145 reproduced orally and very quickly as 3146; 715 also quickly as "714, no 715." The written word *hund* reproduced correctly, but with a capital initial letter.

And being told that they had previously thus reproduced a word in an unknown language (not unknown character), viz., Hungarian, I tried the Greek letters Φαιδω ; this, however, was considered too puzzling and was only reproduced as Uaso. A French high-heeled shoe, made of crockery, being set up as object, it was drawn by the percipient very fairly correctly, and said to be something like a boot, and a protuberance was tacked on where the heel was. (Plate V., R. 12.)

A white plaster cast of a child's hand, next tried, failed to give any impression. An unlighted candle in candlestick was unsuccessful, and it was objected that there was too much glare of light. Subsequently the percipient said she had seen the general outline of a candlestick but did not think of its being the thing. A teapot and a cup both failed, and two of the drawings did not succeed in getting any colourable imitation. Lastly, another set of card trials were made, with the object of testing the effect of various kinds of contact: a card series being quick and easy to run through.

	CARD EXHIBITED TO AGENT.	CARD NAMED BY PERCIPIENT.
Full contact with both hands	Nine	Nine
	King	King
Contact with tips of fingers only	Knave	Two
	Nine	Nine
	Nine	Ten
	Queen	Two
	Eight	Eight
Contact with one finger of one hand	Five	Six
	Seven	Seven
	Three	Four
	Ten	Six
	Queen	Two
	Ace	Ace
No contact	Ace	Four
	Knave	Five
No direct contact, but gap bridged by other person's hand	King	Four
	Four	Eight
	Ten	Seven
Slight contact of knuckles	Eight	Six
	Six	Ace
	Two	Two
Full contact again	Knave	Ace
	Seven	Six
	Three	Three
	Four	Four
	Ace of diamonds	Ace—red—diamond
	Nine of clubs held sideways.	Nine—clubs

The record of this series is more complete than that of another varying contact series, but it did not strike me as so instructive at the time; and as it came toward the end of an evening there was probably some fatigue.

The last two entries represent attempts to get the suit as well; but as the particulars are given in stages there is no particular advantage in thus naming a card completely, and it takes a longer time.

On another evening the amount of contact was varied, but I omitted to call out to the reporter the position of the hands with reference to each other. One hand of each person lay on a table, and I sometimes made them touch, sometimes separated them, all the time going on with the card series. My impression at the time was, as expressed above, that pronounced failure began directly I broke contact, but that mere knuckle contact was sufficient to permit some amount of success.

I can only give the record as it stands. I believe we began without any contact.

Second Card Series. Varying amount of contact: sometimes none.

CARD SHOWN.	CARD GUESSED.
Two of Spades	Knave
Ace of Diamonds	Five
Knave of Diamonds	Knave
10 of Diamonds	9
6 of Hearts	5
8 of Hearts	9
9 of Diamonds	Ace
King of Diamonds	King
10 of Hearts	10
9 of Clubs	9
Ace	Ace
Queen	Two
Queen	Queen
Knave	Ace
King	King
Eight	Eight
Eight	Eight
Seven	Eight
Ace	Ace
Knave	Knave
Seven	Seven
Four	Four
9	6
Queen	3
King	King
Ace	7
Ace	5
5	10

CARD SHOWN.					CARD GUESSED.
5	...	.	.	. .	4
6	...	.		...	7
5	...	.	..	. .	3
6	...	. .	...	...	6
2	...	...	...	. .	3
3	...	...	...	...	6
4	...	...	...	...	4
2	...	...	...	...	8
4	...	...	...	..	5
3	...	...	...	...	4
3	...	...	...	...	Knave

Where lines are drawn it is because I called out some change in the contact; but I made other changes whose occurrence is not recorded.

The only use to be made of the record of this series, therefore, is to treat it as a whole and to observe that out of 39 trials 16 were correct and 23 wrong.

On this occasion there was one reporter who wrote down both what he saw and what he heard; and the operation was so rapid that he had sometimes barely time to do the writing. Towards the end of a series, fatigue on the part of either agent or percipient generally seemed to spoil the conditions.

It is manifest that these experiments should not be conducted too long consecutively, nor repeated without sufficient interval; but if common sense is used there is nothing deleterious in the attempt, and if more persons tried, probably the power would be found more widely distributed than is at present suspected.

I wish to express gratitude to the Fraulein von Lyro and their parents, for the courtesy with which they acquiesced in my request for opportunities of experiment, and for the willingness with which they submitted to dull and irksome conditions, in order to enable me to give as good evidence as possible to the Society.

The illustrations which follow are traced from drawings and reproductions made in the experiments, but though facsimiles in form they do not always represent satisfactorily the relative clearness and faintness of some of the lines. O. in all cases stands for Original drawing, and R. for Reproduction.

V.

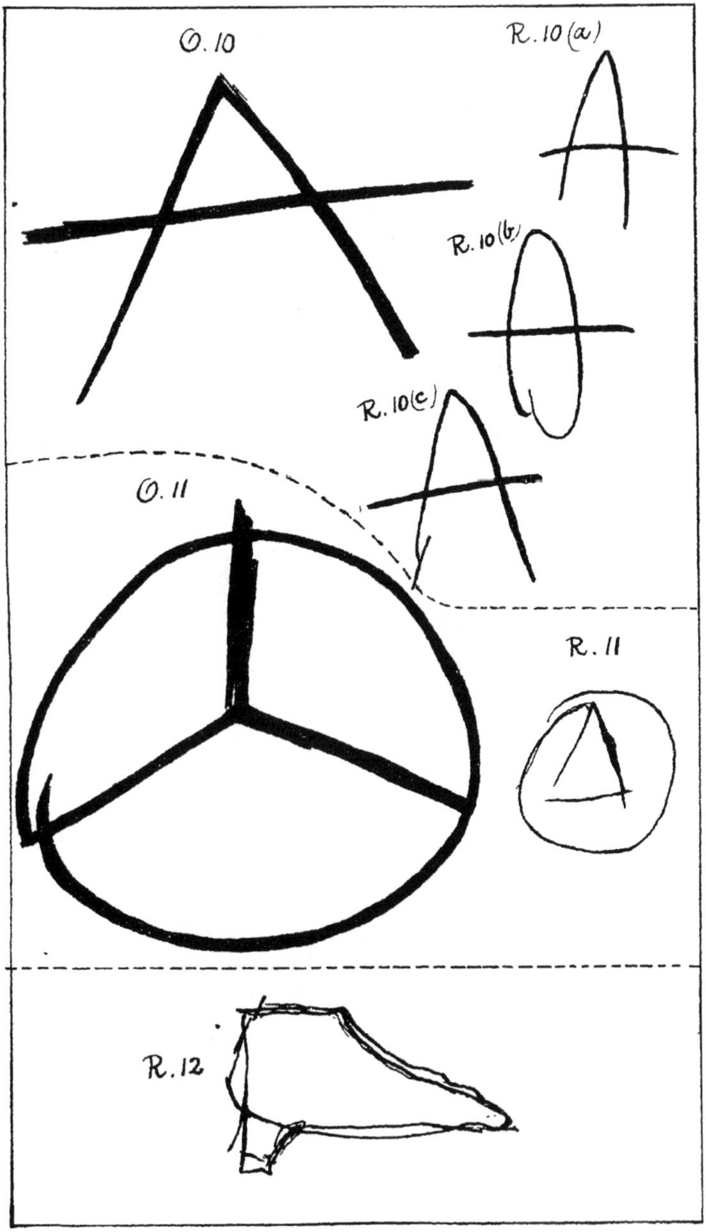

IV

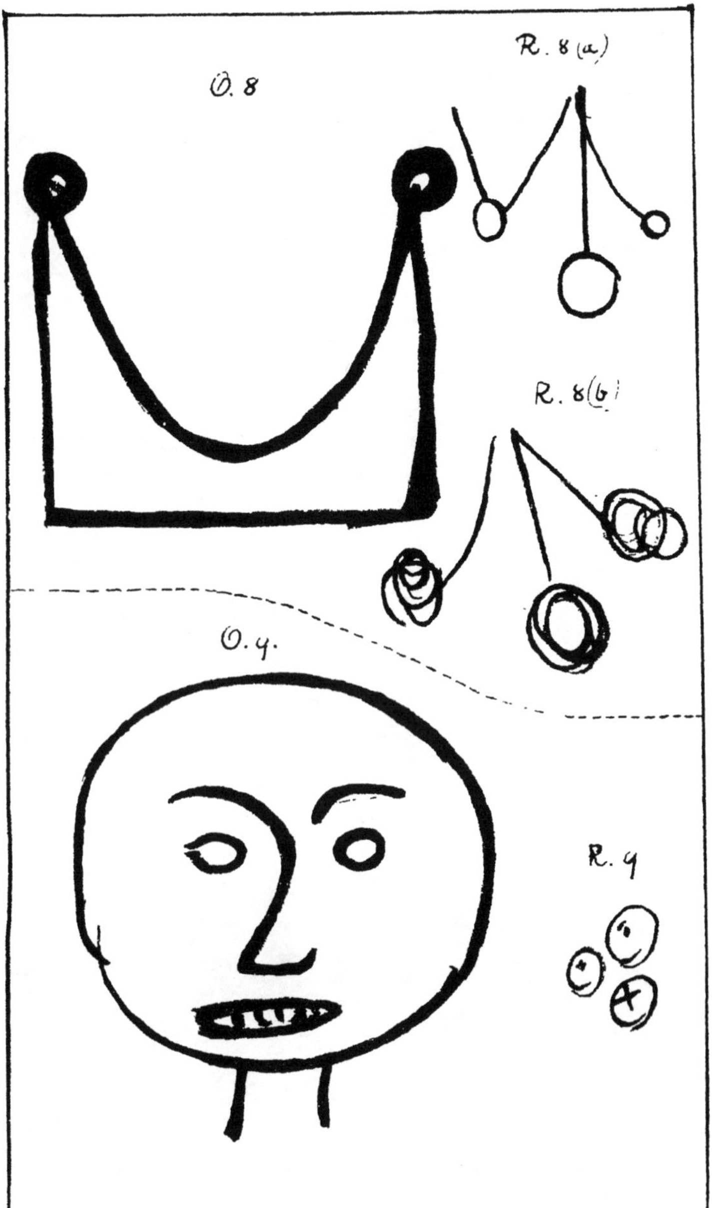

III

O. 6

R. 6

O. 7

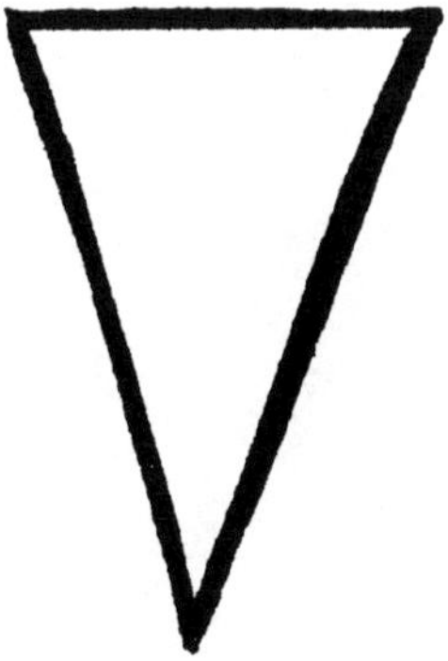

R. 7

II

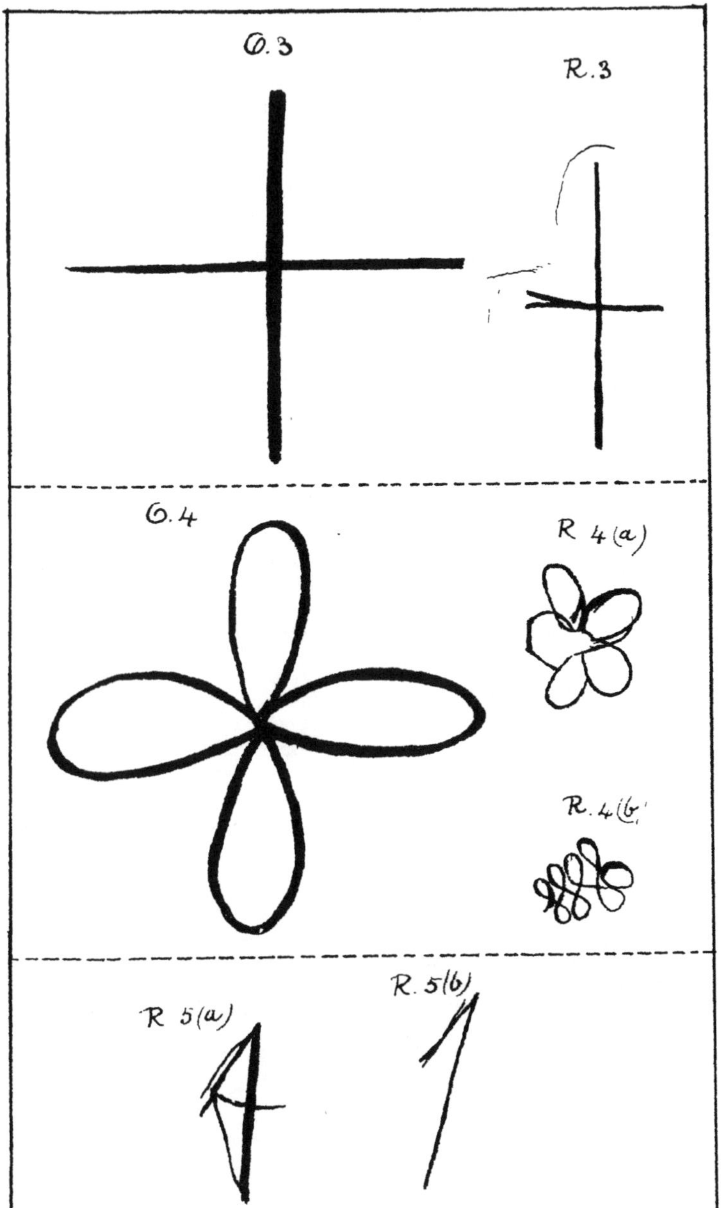

II

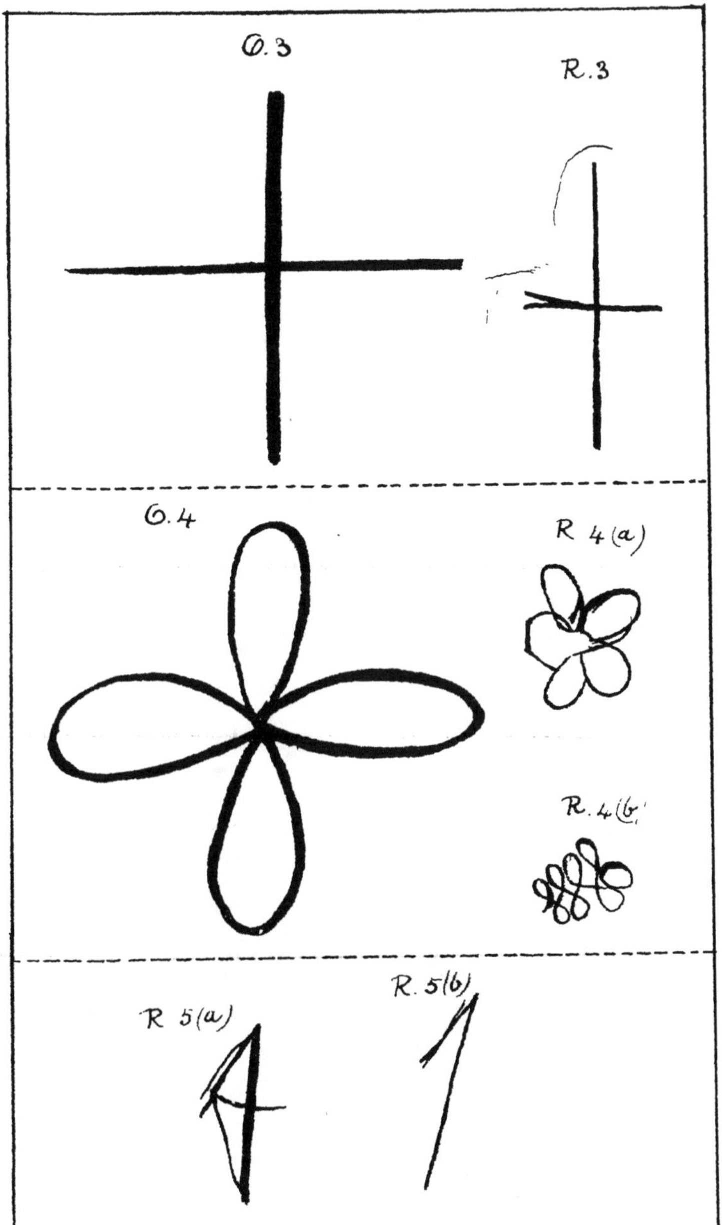

I

R. 1 (a)

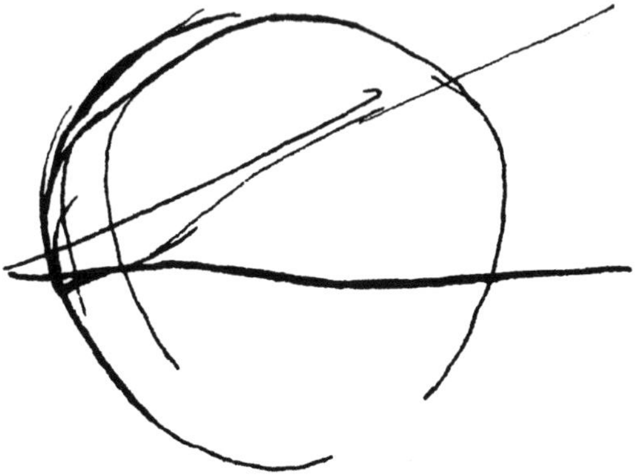

R. 1. (b)

O. 2

R. 2

Made in the USA
Monee, IL
04 November 2026

67388667R00015